The Young Scientist Investigates

Heat

Contents

Fire

No one knows when people first started to use fire. Probably the first fires were started by lightning flashes. The heat of the sun may also have set dried plants alight. Perhaps a brave person managed to get a lighted stick from one of these early fires. He could then have used the lighted stick to start a fire of his own.

People learned how to use fire to cook their food. Fire gave them light and warmth. It could be used to clear the forests to make a place to live and to grow food. And fires kept wild animals away. People later learned how to make fire by rubbing two sticks together. The rubbing made the sticks hot and they started to glow. The people blew on the glowing sticks to make flames.

Other early people discovered how to make sparks with a piece of flint. The sparks could be made to set pieces of dried cloth, bark, or leaves alight. Nowadays we use matches to start a fire.

Bonfires

To build a fire you need wood. You also need some paper and a match. The wood and paper are called fuels. A fuel is something that can burn. Even the match is a fuel. Fuels cannot burn by themselves. If you make a pile of wood and paper, it will not burn. It is not hot enough.

Fuels can burn only when they are hot enough. We often use matches to start a fire. The head of the match contains chemicals. The matchstick usually is made from dry wood. When you rub the match along the side of the box, you make the match head hot. The heat makes the chemicals in the head of the match burn brightly. The fire heats the matchstick, which then begins to burn.

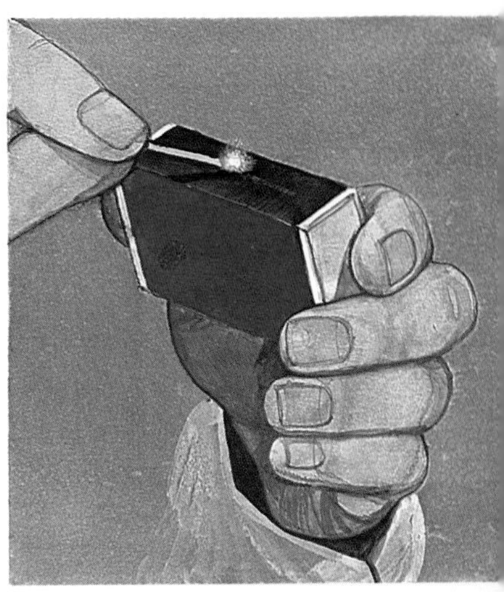

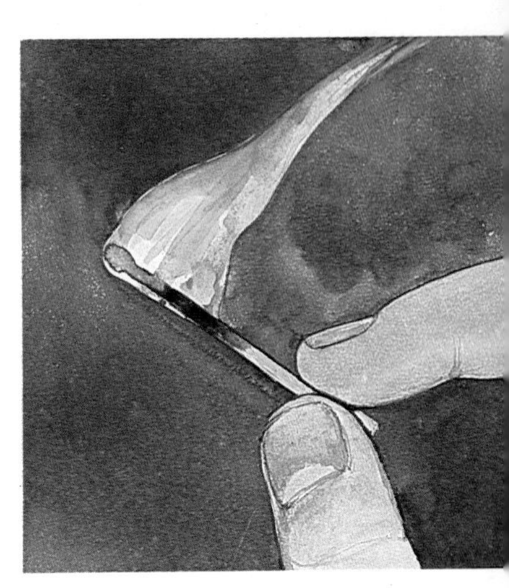

Putting out fires

Putting a jar over a candle

Fire also needs air. Wood, matches, and paper cannot burn without air. If a jar is put over a lighted candle, the flame soon goes out. The candle cannot burn without air. If you make a fire and pack the wood and paper too tightly, they will not burn. The air cannot reach all the fuels easily. The fire begins to smoke and goes out. The gas in the air that is needed to make things burn is oxygen.

Fire can be very useful. But it also can be very dangerous. If you want to put out a fire, you have to stop air from getting to it. Firemen put out most fires with water. Water cools the fire and keeps the air away. Then the fire goes out.

Firemen using water to put out a factory fire

Water cannot put out all fires. Firemen do not use water on burning oil or gas. Oil and gas float on water. The water would spread an oil or gas fire. Firemen spray oil or gas fires with foam. The foam stops air from getting to the fire, and the fire goes out.

Firemen using foam to put out a fire

Fuels

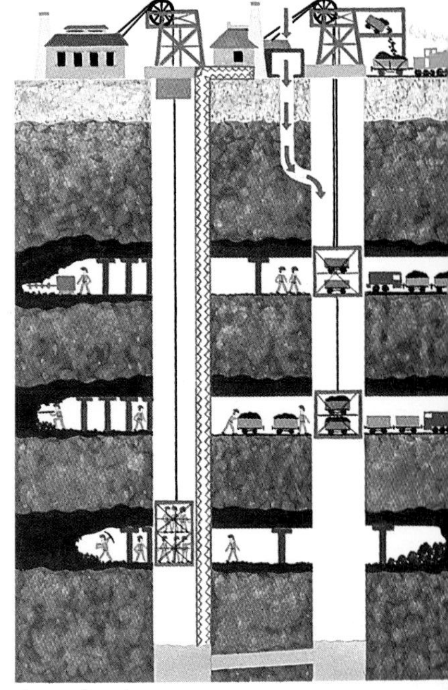
A coal mine

There are many fuels beside wood and paper. Anything that will burn is a fuel. One of the most important fuels is coal.

Coal was made from trees and other plants that grew millions of years ago. The trees and plants died and were covered by mud and sand. The mud and sand pressed down on them. Slowly the trees and plants were turned into coal. Some coal is found near the surface. But most coal is deep underground and has to be mined.

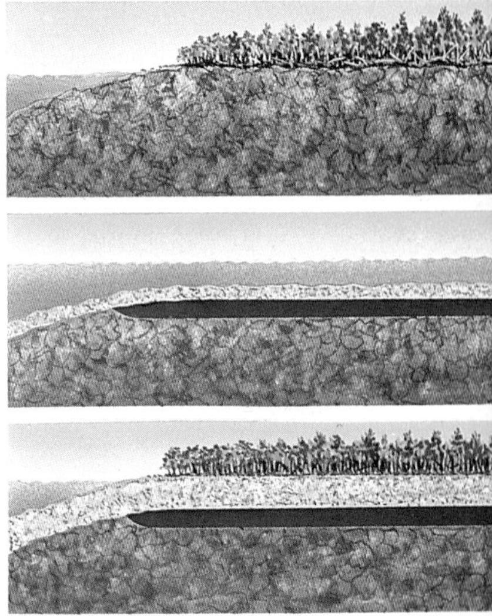
How coal was formed

Another very important fuel is oil. Oil was made from tiny sea plants and animals, which died millions of years ago. The oil is found between layers of rock deep underground. Wells have to be dug to reach the oil.

Gas is also an important fuel. There is gas underground as well as oil. Sometimes oil and gas are found in the rocks under the sea. Floating drilling rigs are built if the sea is deep. Ships or big pipes carry the oil and gas to the land.

The trees and plants that formed coal millions of years ago.

An oil rig

Diagram of an oil well

Small living things like these formed oil

We use fire for cooking. We use the heat we get from fire to warm our houses. Fire can give us light too. Long ago everyone used fire for light. Now most people use electric lights. Electricity is made in a big building called a power plant. In most power plants coal or oil is burned. The burning coal or oil turns water into steam. This steam is used to turn the machines that make electricity.

In the engine of a car, gas is burned. Gas is made from oil. The burning gas turns the engine. This makes the wheels go round. Burning oil makes truck and tractor engines work. Fires burning in the engines of airplanes help them fly. Many trains are powered by burning oil in their engines.

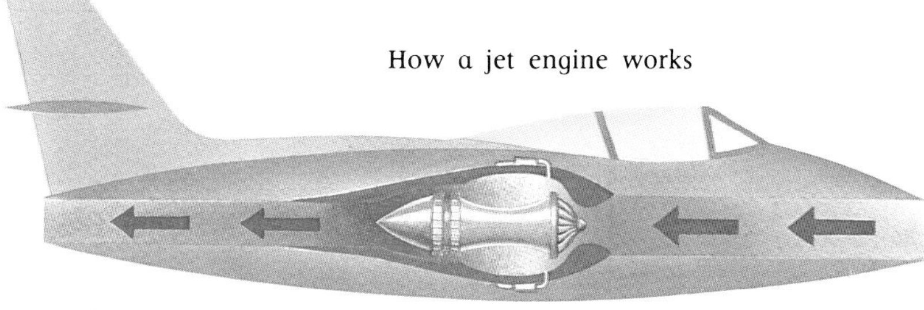
How a jet engine works

Diagram of a car engine

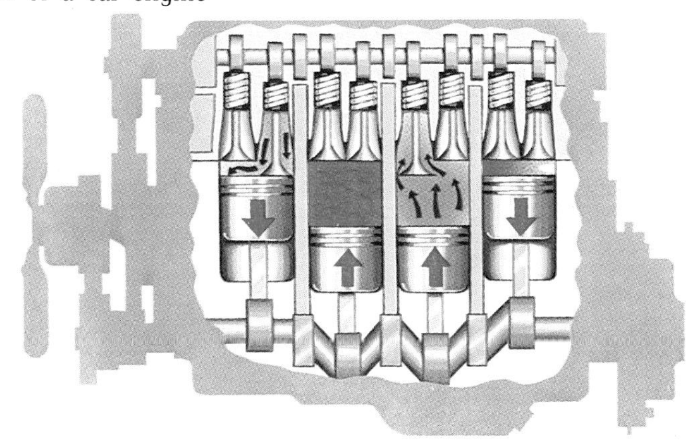

Getting warm

To get warm we may sit in the sun or in front of a fire. We can feel the rays of heat coming from the sun or from a fire. Sometimes the room is heated by a fire that is in another part of the building. Usually the fire is in a boiler. Boilers may burn coal, oil, gas, or some other fuel. The boiler heats water. The hot water passes along pipes into radiators in the room. The hot radiators warm the room. After the water has been through the radiators, it is no longer as hot as it was originally. So it goes back to the boiler to be heated again.

There are many other ways in which we get warm. We get warm if we walk fast, run, or jump about. This is because some of our food is used up inside us, to make heat. The food makes heat in our muscles. We are really burning up the food, although there are no flames or smoke. If we are too hot we sweat. Sweating helps to cool us.

Diagram of a house central heating system (simplified)

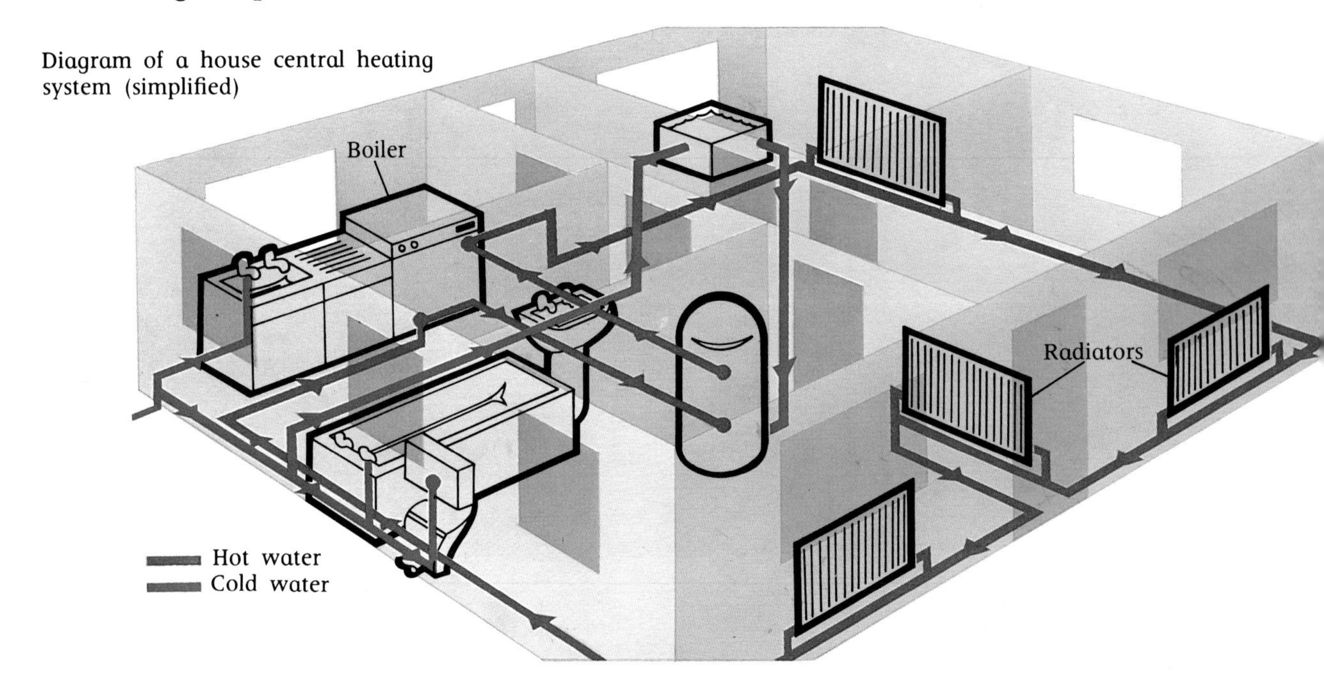

Keeping warm

Our clothes help keep us warm. In cold weather we wear extra clothes. Our clothes stop our bodies from losing heat to the moving air around us. Anything that is hot loses heat to the cooler air around it. A hot teapot soon becomes cold. To keep the tea hot longer, we put a cozy on the teapot. The cozy stops the air from moving around the teapot and making the tea cold.

A tea cozy

Buildings can lose heat to the air in cold weather. Much of the heat from fires and radiators is lost through windows, walls, and roofs. Cold moving air, called a draft, also comes into buildings through cracks around the door and window frames. To keep the heat in a building it is necessary to stop drafts. This can be done by sealing up the cracks around the doors and window frames. The roof of the building and the gap in the middle of the walls can be lined with a warm material to keep the heat in.

The inside of a roof lined to keep the heat in

Some windows are double glazed. They have two sheets of glass with a gap in between. The air in the gap is still. It does not let the heat escape through the window so easily.

Ways in which a house can lose heat to the air in cold weather

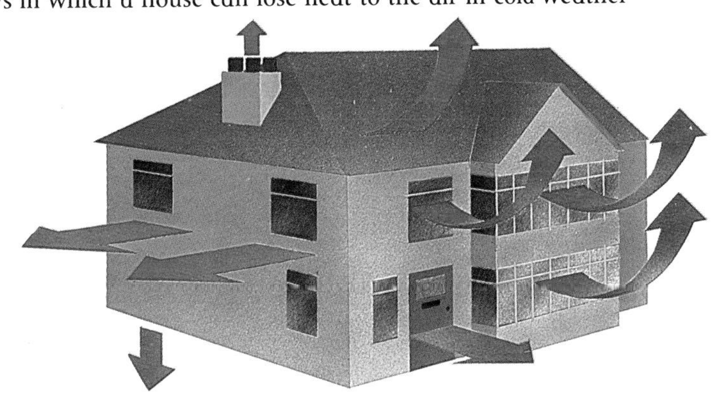

Melting things

If a solid substance is heated enough, it will turn into a liquid. We say the solid has melted. Ice on puddles soon melts when the sun comes out. The solid ice turns to liquid water. A Popsicle quickly melts on a hot day. When we light a candle, some of the wax melts as the candle burns.

Even rocks can melt if they are heated enough.

Deep inside the Earth it is so hot that the rock is a liquid. We can see this liquid rock, or lava, when a volcano erupts. Most of the metals we use are found in rocks. Rocks with a lot of metal in them are called ores. The ore must be burned in a big fire, or furnace, to get the metal out.

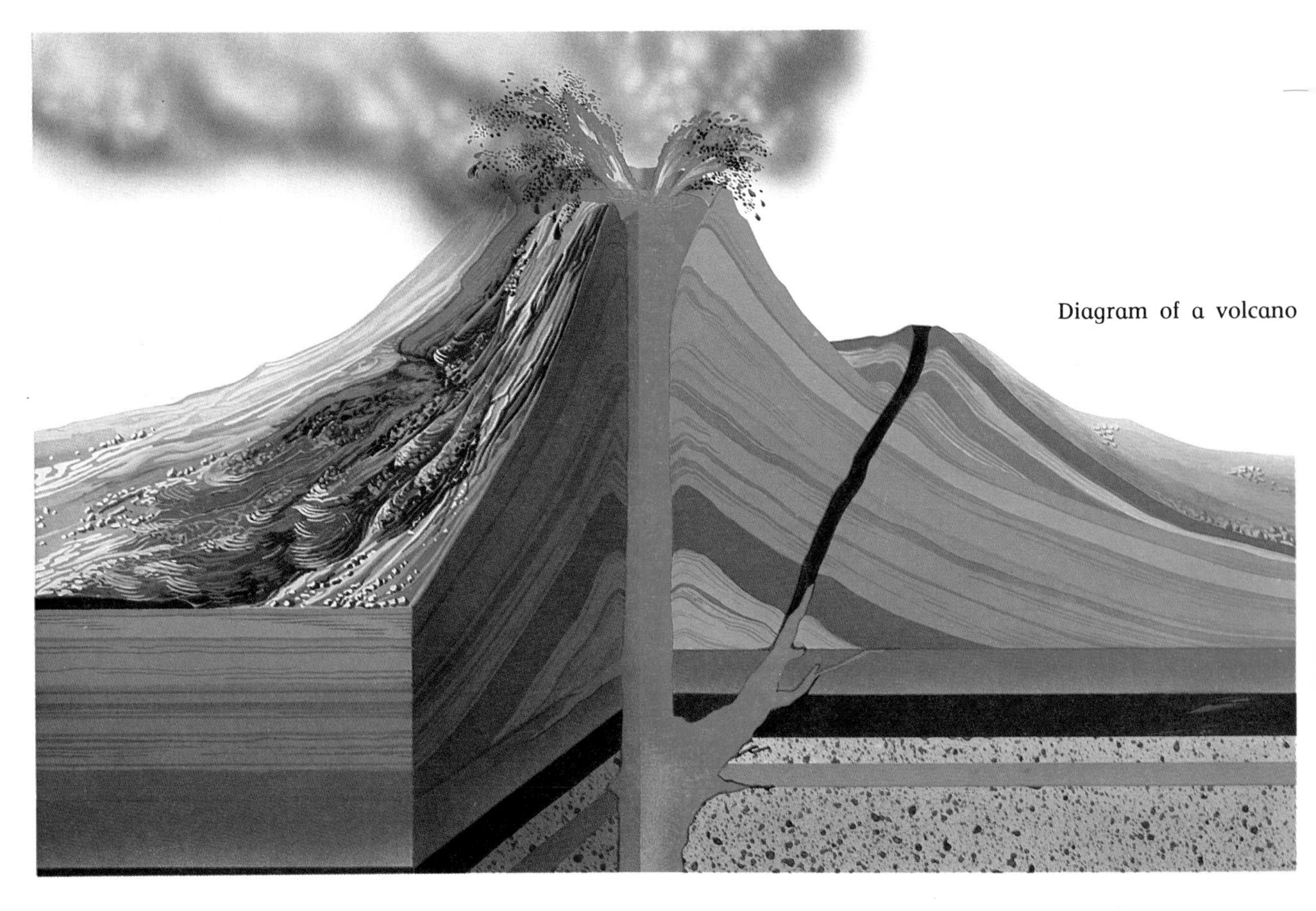

Diagram of a volcano

Digging iron ore

To make iron, the iron ore is heated in a big furnace. The iron ore melts. Liquid iron sinks to the bottom of the furnace. The liquid iron is taken from the bottom of the furnace. When it cools it forms solid iron. If any liquid is cooled enough, it will become a solid. If water is cooled it becomes solid ice. When the lava from a volcano cools, it forms solid rock.

Liquid iron being taken from a furnace

Do you remember?

(Look for the answers in the part of the book you have just been reading if you do not know them.)

1 What did early man use fire for?

2 How did early man make fire?

3 What do we call something such as wood, paper, and matches that can burn?

4 What has to happen before something will burn?

5 What are the head and stick of a match made from?

6 What happens when you rub the head of a match along the side of the box?

7 Which part of the air is needed for things to burn?

8 What must you do to put out a fire?

9 Why don't firemen use water to put out oil and gas fires?

10 What was coal made from?

11 What was oil made from?

12 Name some of the ways in which fire helps us to travel from one place to another.

13 What do power plants use burning oil or coal for?

14 How does our food make us warm?

15 Name some other ways in which we warm ourselves.

16 How do our clothes stop us from losing heat?

17 What does a tea cozy do when it is put on a hot teapot?

18 What can be done to buildings to stop them from losing the heat inside?

19 What happens to a solid substance such as rock if it is heated enough?

20 How is iron made from iron ore?

Things to do

1 **Make a picture.** Show a caveman lighting a fire by rubbing two sticks together. Put the man's wife and children in the picture as well. What kinds of clothes do you think they would be wearing? Would there be any animals around the cave? If so, put these in your picture.

2 **Rub your hands together hard.** What do you feel? Use a saw to cut a piece of wood. Stop sawing, and then carefully touch the blade of the saw. What do you feel? Rub a long nail backward and

forward along an old brick. Feel the end of the nail. What do you notice?

What have you learned about rubbing things?

3 Write a poem about a bonfire. Here are some words you might use in your poem:

fuel	smoke	sparks	ashes
wood	smolder	flames	flickering
crackle			

Make some music to go with your poem. What instruments will you use for the bonfire? Write down your music if you can.

4 Make a collection of pictures showing fires. Write a sentence or two about each of your pictures.

5 Find out all you can about firemen. Collect pictures of firemen. Find out what kinds of protective clothes they wear and what kinds of equipment they use. Collect pictures of firemen and fire engines. Make a book about firemen using your pictures and the information you have collected.

6 Use your imagination. Have you ever looked into a fire and imagined that you could see faces, animals, birds, or other shapes? There is an old story about a bird called the phoenix. There was only one phoenix and it was supposed to live in the desert for 500 or 600 years and then burn itself to death in a fire. A new, young phoenix was then believed to arise from the ashes of the old.

Write your own story about the phoenix, or make up a story about an animal, bird, or person who lives in a large fire.

7 Pretend that you have a time machine that will allow you to go back into the past. Pretend that you go back thousands of years to the time when trees and other plants are dying and forming coal. Write a story about a day in your life. What is the countryside around you like? What wild animals are there? What do you eat? How do you collect or catch your food?

8 Find out how your school is kept warm. If possible, ask your teacher if you can see the boiler room. Perhaps your school janitor will be able to show you how the boiler works. What kind of fuel does it use?

Make a plan or drawing showing how the pipes carry the hot water from the boiler to the radiators in the different classrooms.

9 Make a list or collect pictures of things that change when they are heated. You might start with coal, wood, water, bread, and an egg. How many more can you think of? How are these things changed by being heated?

10 Make a scrapbook of pictures of clothes. Divide the pictures into four big groups: clothes worn by boys, girls, men, and women. Then divide each of the four groups of pictures into those of clothes worn in summer and those worn in winter.

Make another collection of pictures showing the clothes boys, girls, men, and women wear in very hot countries and in very cold countries.

Expansion and contraction

Have you ever looked at telephone wires on a hot day? On a hot day the wires look like this.

But on a cold day, the telephone wires are tight like this.

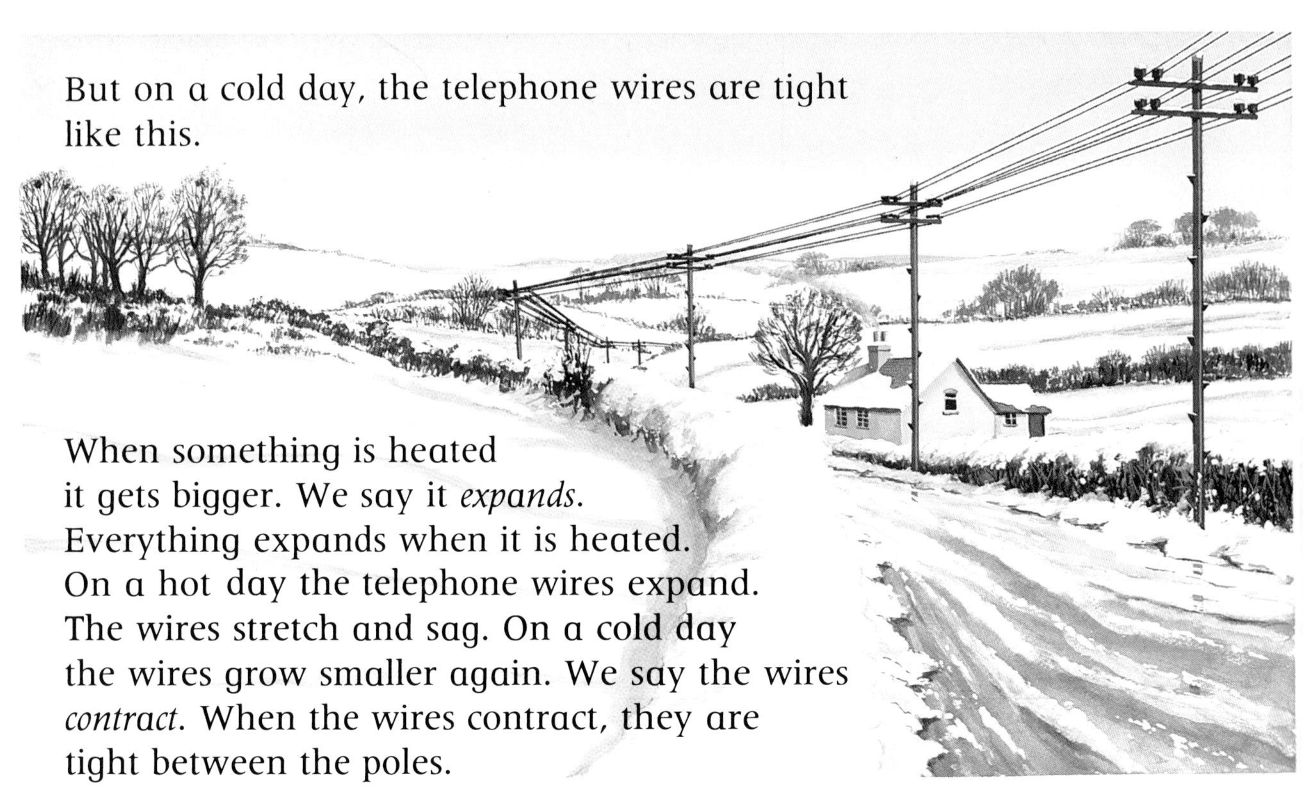

When something is heated it gets bigger. We say it *expands*. Everything expands when it is heated. On a hot day the telephone wires expand. The wires stretch and sag. On a cold day the wires grow smaller again. We say the wires *contract*. When the wires contract, they are tight between the poles.

Little gaps are left between the sections of a highway bridge. The gaps are left so that the bridge can expand when the weather is hot. If the gaps were not left, the bridge might expand and buckle up.

A metal screw top that is stuck on a bottle can be removed by running hot water over it for a few minutes. The bottle top expands in the hot water. Then the top comes off the bottle easily.

Temperature

This is a thermometer. The thermometer measures the hotness or coldness of the room. The liquid in the thermometer expands when it is warmed. When it is cold the liquid contracts. And so the level of the liquid in the thermometer changes with the temperature of the room. The temperature on the thermometer is shown in degrees. Each mark on the thermometer is one degree. There are 100 marks on this thermometer. It is called a centigrade, or celsius, thermometer (cent means 100). With this thermometer the temperature of boiling water is 100 degrees centigrade (100°C). Water freezes at 0°C on this thermometer.

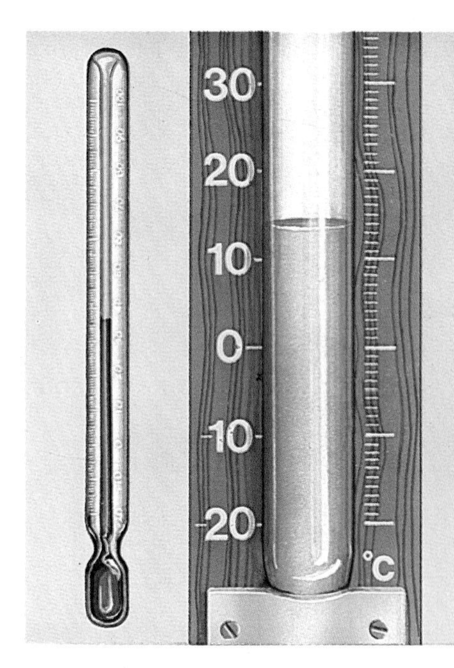

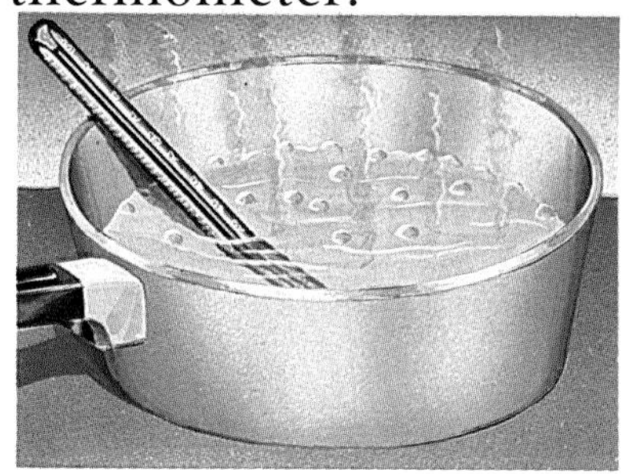

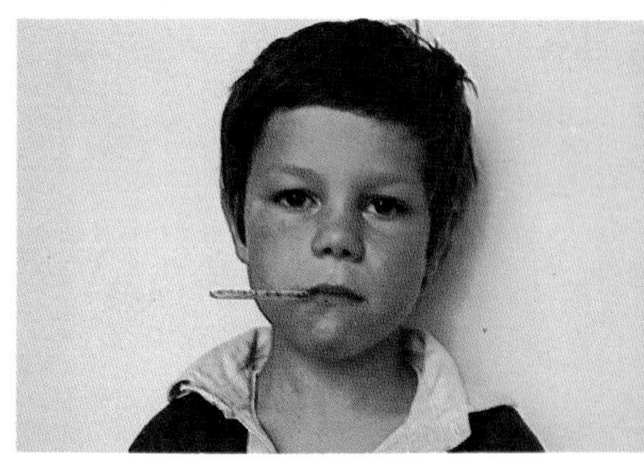

This is another kind of thermometer. It is used by the doctor to take your temperature if you are ill. This thermometer tells you the temperature of your body. The doctor puts the thermometer into your mouth for about two minutes. He takes the thermometer out and reads what it says. Your temperature will be about 98.6°F. when you are well. It will be higher than this if you are ill. Your temperature stays the same whether the day is very hot (100°F.) or very cold (33.8°F.). Your temperature stays the same whether you are in a hot bath (104°F.) or in a cold one (50°F.).

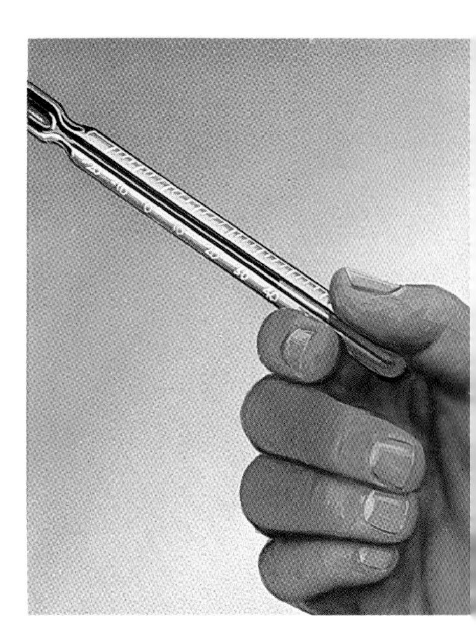

Animals and temperature

Some warm-blooded mammals

You are warm-blooded. It does not matter if the weather is warm or cold, your temperature will stay the same. All kinds of birds are warm-blooded. So are all the animals we call mammals. Mammals have hair or fur on their bodies and feed their young on milk. Dogs, cats, rabbits, mice, cows, sheep, horses, and humans are all mammals. Their temperature stays the same whatever the weather. These warm-blooded animals can be active whether it is hot or cold.

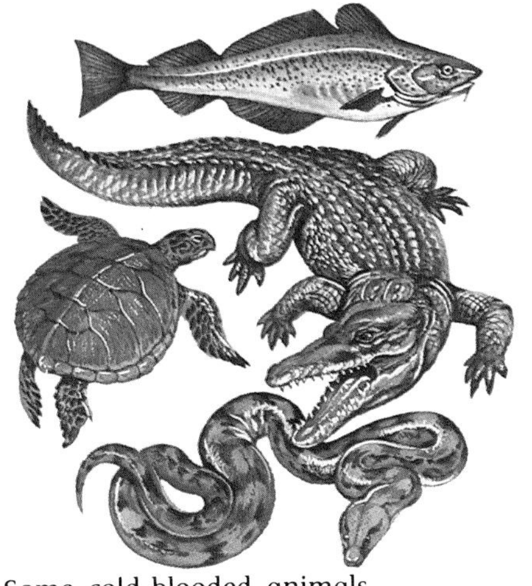
Some cold-blooded animals

Many animals are cold-blooded. Fish, snakes, lizards, tortoises, frogs, toads, newts, insects, and spiders are some cold-blooded animals. The temperature of these animals is the same as that of the air or water around them. Cold-blooded animals can be active only when the air or water is warm. When the air or water is cold, cold-blooded animals sleep through the cold weather. Animals that sleep through the winter are said to hibernate.

A hibernating queen wasp

A toad hibernating
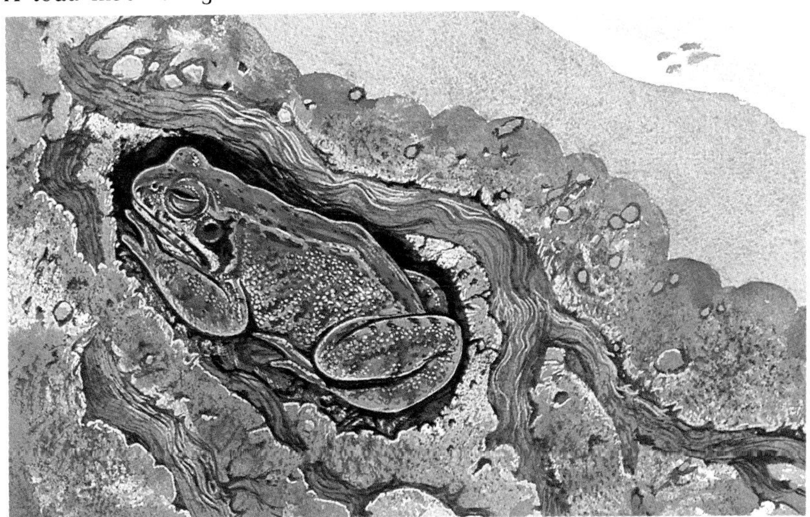

Sunshine

The boy in the picture is warming his hands in front of an electric heater. Rays of heat travel from the heater in a straight line. If someone puts a screen in front of the heater, the boy will not feel the heat. This is because the screen stops most of the rays of heat. If the screen is shiny or light colored, most of the rays of heat bounce back toward the heater. We say the rays are *reflected.*

Many electric heaters have shiny metal behind them. The shiny metal reflects the rays of heat into the room. The sun's heat also reaches us in rays. The sun's rays travel in straight lines to reach us. Because the sun's rays travel in straight lines, you can stay cool in the shade of a tree. When there are clouds in the sky, we do not feel as warm. The tiny drops of water in the clouds reflect the sun's rays.

White clothes reflect the sun's rays. White clothes are cool in summer.

Dark colors soak up, or absorb, the sun's rays. Dark-colored soils are warmer than light-colored soils and grow crops quicker. Dark-colored clothes absorb heat from the sun. They can be very hot in summer.

How clouds reflect the sun's rays

People in hot countries often wear light-colored clothes. Why?

Expanding air

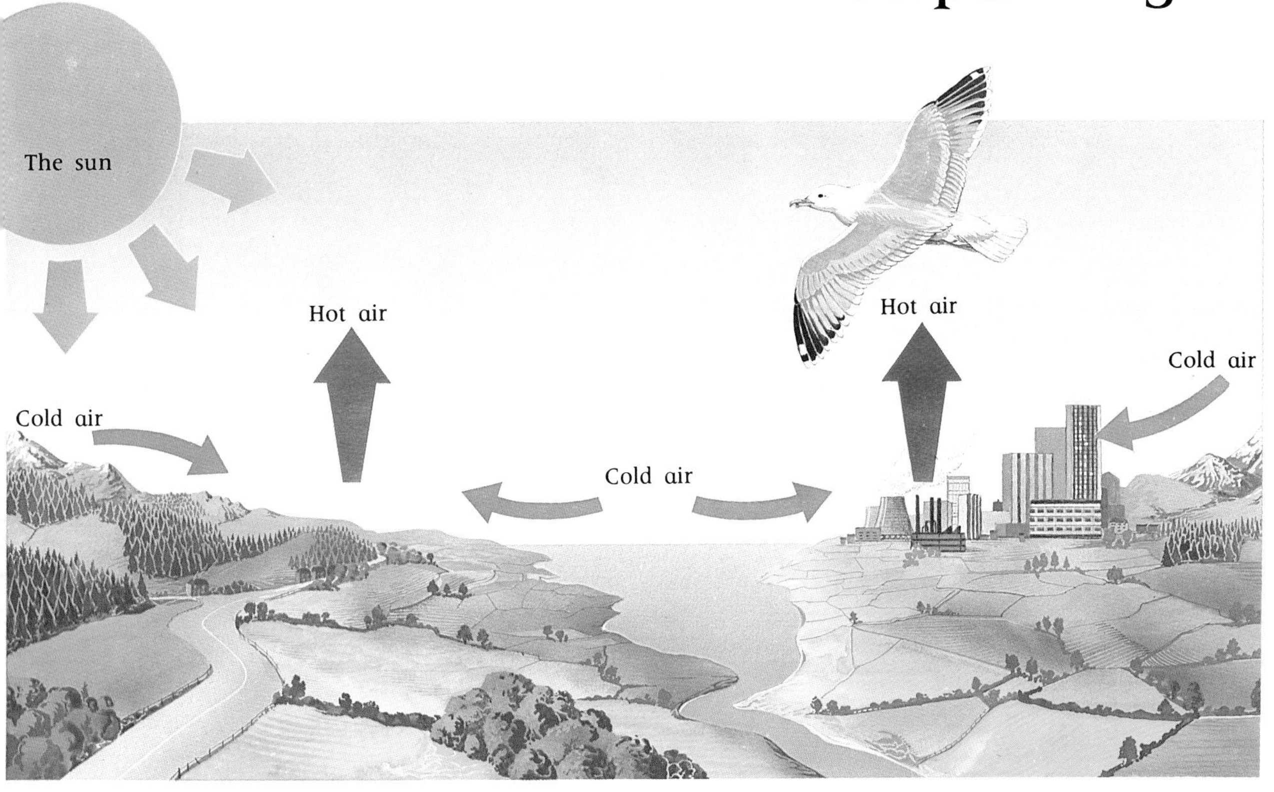

Hot-air balloon

On a warm day the sun heats up the land. The hot land heats the air above it. The warm air rises. Cooler, heavier air moves in to take its place. This is happening all the time somewhere in the world. We call these movements of cooler air, wind.

If you hold your hand above a hot radiator, you can feel the hot air rising. The hot air rises because it has expanded. It is lighter than the cold air around it. Hot-air balloons work because the air in them expands when it is heated. A hot-air balloon is made of very light material. The balloon has an open neck. A gas flame is used to heat the air inside the balloon. Once the air is heated, it expands. The balloon is quickly filled with hot expanded air. This hot air is lighter than the cold air outside. And so the balloon soars up into the sky.

Water vapor

If a liquid such as water is heated, it gets warmer. The temperature of the water rises. The water turns into a gas called water vapor, which disappears into the air. We say the water has evaporated.

Boiling water in a kettle

Water heated in a kettle turns to water vapor. When the vapor cools it turns back to water. Water from a boiling kettle turns back to drops of water that float in the air. We call these tiny drops of water, which float in the air, steam.

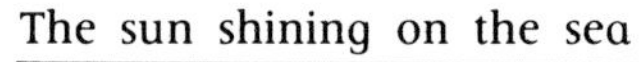
The sun shining on the sea

A snow scene

The sun shining on the sea turns a lot of water to water vapor all the time. The sea water evaporates and forms water vapor. The water vapor disappears into the air. High in the sky it is cold, and the water vapor cools to form clouds. Clouds are made of millions of little drops of water that float in the air. If the clouds are cooled still more, the tiny drops of water join together. The big drops that are formed are too heavy to float in the air. They fall to the ground as rain. When the weather is very cold, the tiny drops of water in the clouds may turn to ice. Each little piece of ice forms a shape called a crystal. The ice crystals grow bigger and fall as snowflakes.

A snowflak

The sun
Clouds
Clouds rise
Rain
Rain and snow
Water vapor
Water vapor
Sea
Water vapor
Lake
River

Conductors and insulators

Some materials allow heat to pass through them very easily. Metals allow heat to pass through them easily. That is why saucepans and kettles are made of metal. Radiators are also made of metal. Because they let heat pass through them easily, we say metals are good *conductors* of heat.

Metals feel cold to the touch. That is because they carry the heat away from your fingers quickly. Some other materials feel warm to the touch. Wood, plastic, cloth, and wool feel warm to the touch. Wood, plastic, cloth, and wool do not carry the heat away from your fingers quickly. They are said to be *insulators.*

We use insulators for oven gloves and for the handles of saucepans, irons, and kettles. That way we do not burn our fingers. We use insulators for table mats. Then we do not mark the furniture with hot plates and dishes. We use insulators around the hot-water tanks in our houses. Then the hot water inside does not cool down as quickly. We use insulators in the walls and roofs of houses. Then the houses do not lose heat as quickly.

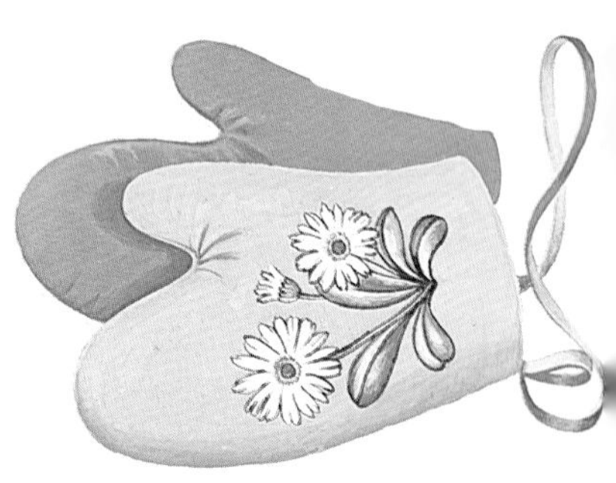

An insulated hot-water tank

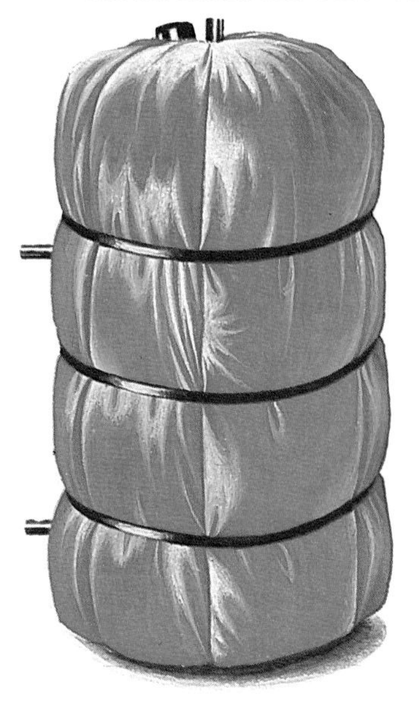

Air as an insulator

Robin in cold weather

obin in warm weather

If air is moving it quickly carries heat away and cools things. We soon feel cold if we stand around when a breeze is blowing. But still air is a good insulator. Things that trap air and keep it still are good insulators.

In cold weather birds fluff out their feathers. The feathers trap lots of air and keep the birds warm. In hot weather, the birds squash their feathers down. The feathers do not trap as much air. And so the birds stay cool.

'hale

Mammals trap air in their fur or hair in cold weather. Then the hair or fur keeps them warm. Many animals grow thicker fur for the winter. We humans do not have much hair on our bodies. But underneath our skin we have a layer of fat. Fat is a good insulator. It keeps us warm in cold weather. Whales, seals, walruses, and penguins have a very thick layer of fat under their skins. The fat helps to keep them warm in the very cold waters where they live.

Penguins

Walrus

Seal

Do you remember?

1 What happens to the telephone wires on a hot day and on a cold day?

2 What do we say happens when something gets bigger when it is heated?

3 What do we say happens when something gets smaller when it is cooled?

4 Why are little gaps left between the sections on a highway bridge?

5 How could you get a metal top off the bottle on which it has become stuck?

6 What does a thermometer do?

7 What are the numbers on a thermometer called?

8 At what temperature does water boil and freeze on a Centigrade (Celsius) thermometer?

9 What is your temperature when you are well?

10 Name six warm-blooded animals.

11 Name six cold-blooded animals.

12 What do many cold-blooded animals do in the winter?

13 Why do many electric heaters have shiny reflectors behind them?

14 Why do we not feel so warm when there are clouds in the sky?

15 What do darker colored soils and clothes do to the sun's rays?

16 What does air do when it is heated?

17 How does a hot-air balloon work?

18 What does a liquid such as water do if we heat it?

19 How is snow formed?

20 How do we describe substances such as metals that allow heat to pass through them easily?

21 How do we describe substances such as wood, plastics, and cloth that do not allow heat to pass through them easily?

22 Why are the handles of saucepans, kettles, and irons often made of plastic?

23 What does a bird do with its feathers in cold weather?

24 What does the layer of fat underneath our skin do?

Things to do

Ask your teacher to let you look at a thermometer. Handle it gently. It will break if you are not very careful. Use your thermometer to find the temperature of things. You might try the temperature of the skin of your fingers, water from a tap, a cup of tea, some ice or snow, the emperature in the middle of the playground on a sunny day and under the shade of a tree on the same day, and the emperature of the classroom. Ask your eacher to find the temperature of boiling water for you.

Take the temperature in the same spot in the playground at the same time every day. A good time would be 9 o'clock in the morning.

Make a graph showing how the emperature changes from day to day.

Hang a thermometer on a wall outside he school. Take the temperature every 30 minutes or every hour throughout one whole day.

Make a graph showing how the emperature changes. When is the hottest ime of the day and when is it coldest?

2 Collect pictures or write down the names of three kinds of furry animals (mammals) that hibernate. Why do you hink they hibernate when furry animals are warm-blooded?

3 Collect pictures to make two wall charts or scrapbooks. One could deal with warm-blooded animals and the other with cold-blooded animals. Write two or three sentences about each of your pictures.

4 Which birds fly south when the days become shorter and colder? Make a list of them. Give another reason why they fly south. Where do they go?

5 Make paper snowflakes. Take a sheet of white or silver paper. Lay a saucer on it and cut around it. Fold the circle of paper into six and cut pieces out of it. When you unfold the paper it will look like the snowflake on page 20. Mount your snowflake on sheets of black paper.

6 Make some Popsicles. Take the ice cube tray from the refrigerator. Make some fruit drinks by adding fruit punch or lemonade powder to water in a glass or jug. Fill the compartments of the ice cube tray to the top with fruit drink. Stand a clean Popsicle stick in each compartment. Put the ice cube tray back in the freezer. How long do the Popsicles take to form? What mixture of water and fruit punch or lemonade powder gives the best tasting Popsicles?

Experiments to try

7 Keep a weather notebook. Every day write down what the weather is like and what the temperature is. Look at the temperature at the same time each day.

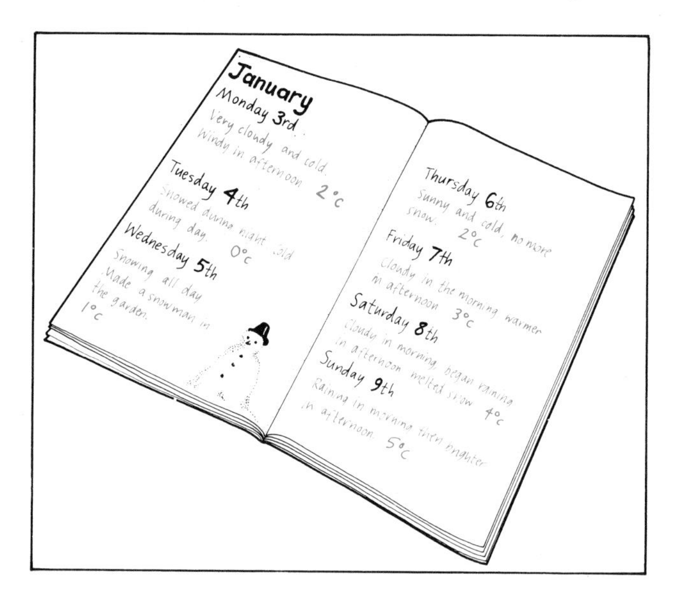

8 Find out what kinds of houses or shelters people build in different parts of the world. Collect pictures of these houses or shelters. Do the people build them to keep warm, to keep out the rain or sun, or to protect them from wild animals?

What kinds of shelters do people make for horses, cows, sheep, pigs, and chickens?

9 Make a wall chart showing animals in winter. Write a sentence or two about each animal.

10 Find out all you can about the mammals and birds that live in cold countries. Collect pictures of the animals. Find out how they keep warm. You might also collect pictures of mammals and birds that live in very hot countries. How do these mammals and birds keep cool?

Do your experiments carefully. Write or draw what you have done and what happens. Say what you have learned. Compare your findings with those of your friends.

The hot water you use in these experiments should be just hot enough for you to be able to put your hand in it. ***Do Not Use Boiling Water.*** Ask your teacher to help you with the experiments where hot water is needed.

1 Can we make a mistake about what is hot and what is cold?

What you need: Three bowls.

What you do: Fill two of the bowls with water. In the first bowl put water as hot as you can stand with your bare hand. **(Careful!)** In another bowl put cold water from the tap.

Into the third bowl put an equal amount of hot and cold water from the other two bowls. This will give you a bowl of warm or tepid water.

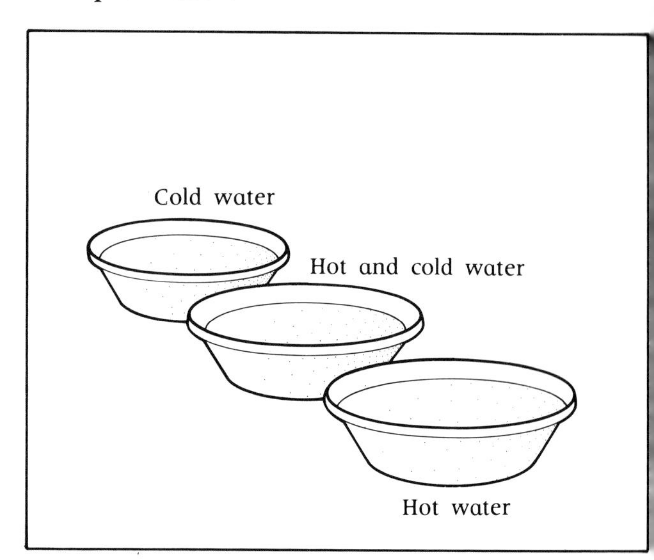

Put the bowls of water on the table. Put the bowl of hot water on the right, the warm water in the middle, and the bowl of cold water on your left.

Hold your right hand in the bowl of hot water and your left hand in the cold water. How does each hand feel?

Count to 20 slowly and then move your left hand from the cold water to the warm water. How does your left hand feel now?

Keep your left hand in the middle bowl and now put your right hand in the middle bowl too. How does your right hand feel now? Can you always tell the difference between hot and cold?

2 What things does heat go through?

What you need: A rubber hot-water bottle; a piece of wood; a plastic plate; a ceramic plate; a flat piece of metal; some cardboard and paper, pieces of wool; and cotton materials.

What you do: Ask your teacher to fill the hot-water bottle with hot water for you. Lay the hot-water bottle on the table. Quickly touch the outside of the bottle. Is it hot?

Now lay the piece of wood on top of the bottle. Leave it there for a few minutes. Put your hand on the top of the piece of wood. Does it feel hot? Does it feel warm?

Next lay the piece of metal on top of the hot-water bottle. Leave it for a few minutes. Now put your hand on the metal. Is it hotter or colder than the wood was?

Now try the experiment with the other things. Make lists of those materials that let the heat through quickly and those that let heat through slowly. Which of the substances are heat conductors and which are insulators?

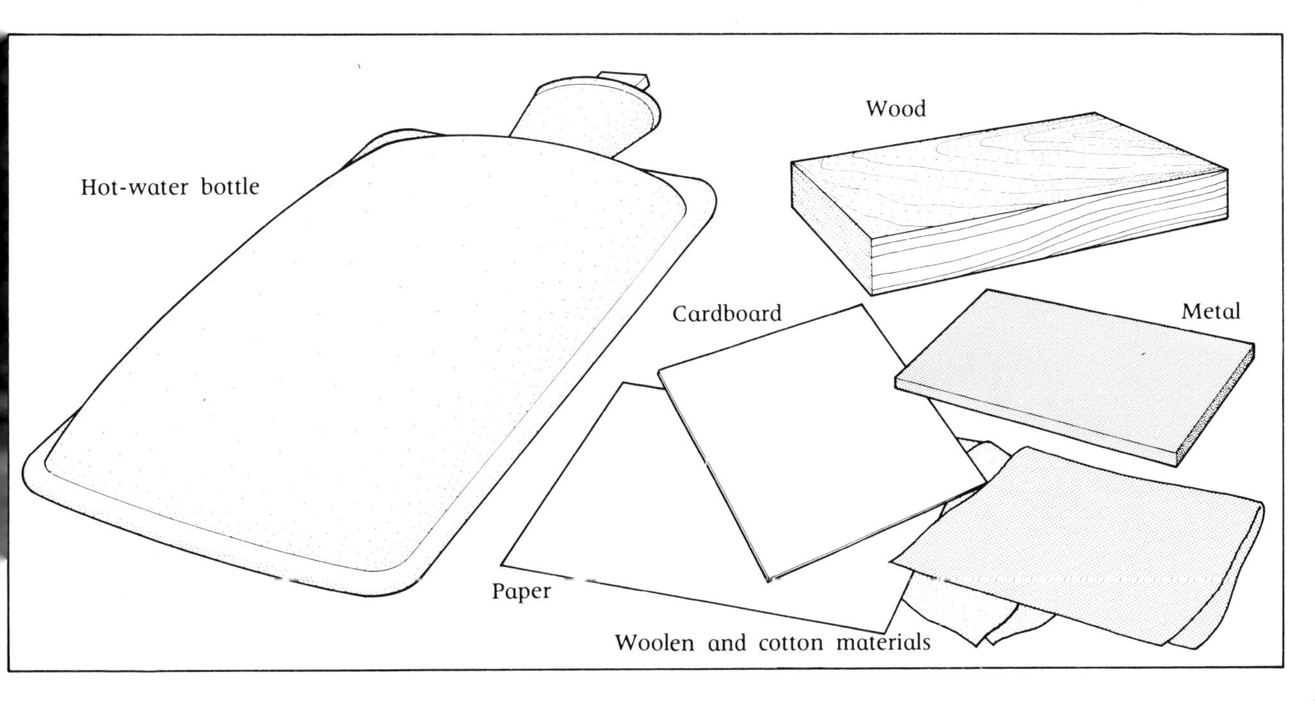

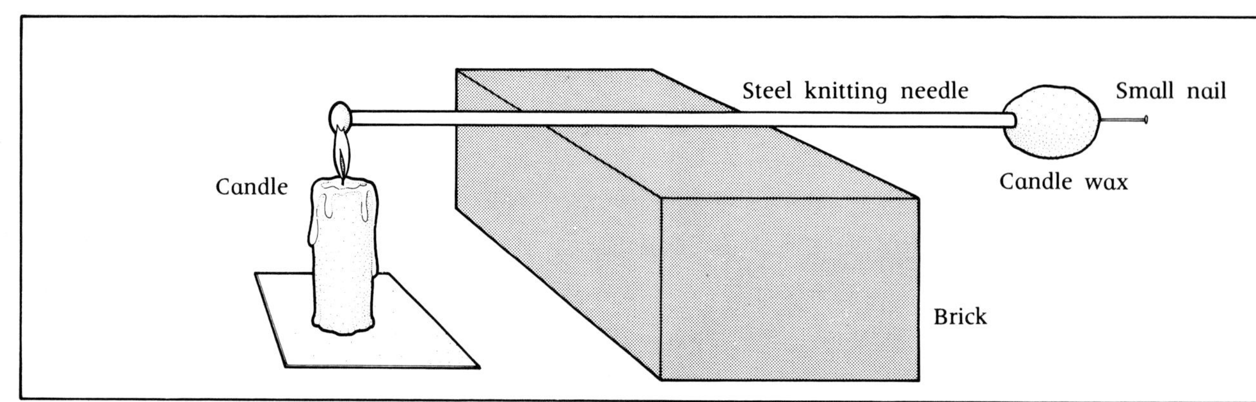

3 How quickly does heat travel along a metal knitting needle?

What you need: A steel knitting needle; a brick; a candle; some candle wax; a small nail.

What you do: Ask your teacher to help you with this experiment.

Roll the candle wax in your fingers until it is soft. Then push the wax onto the end of the knitting needle. Stick the small nail in the wax. Rest the needle on the brick as shown in the picture above. Ask your teacher to light the candle, and then use it to heat the end of the needle away from the wax.

What happens? Does the heat travel along the needle? If so, how long does it take to reach the other end of the needle? What happens to the small nail? What happens to the wax?

4 What happens when air is heated?

What you need: A tin with a lid (a clean cocoa or chocolate syrup tin will do); a bowl; a jug.

What you do: Put the lid on the tin. Do not push it on too hard. Stand the tin in the bottom of the bowl. Now fill the jug with hot water from the tap and slowly pour it **(Careful!)** into the bowl.

Watch the lid of the tin. What happens to the lid? Why do you think this happens?

5 Another experiment to see what happens when air is heated

What you need: A plastic bottle; a plastic bucket; a balloon.

What you do: Cool the plastic bottle by standing it in cold water or in the refrigerator. Stretch the neck of the balloon and fit the balloon over the neck of the bottle in the way shown in the picture.

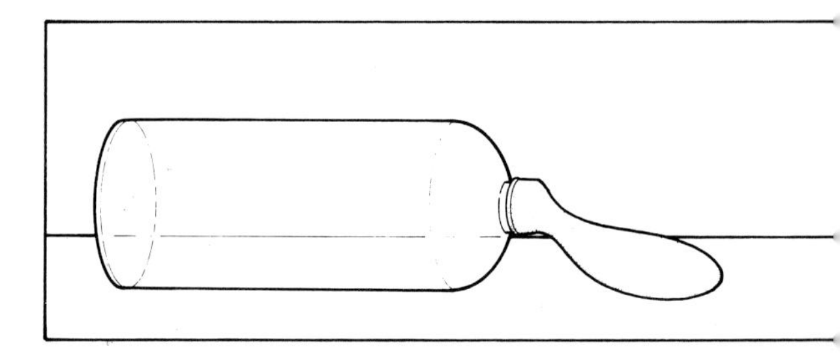

Now place the bottle and balloon in a bucket of hot (*not boiling*) water. This will warm the bottle and the air inside it.

What happens? Why do you think this happens?

How does a greenhouse work?

lany gardeners use a greenhouse to make ıeir plants grow quicker. In this xperiment we shall see how a greenhouse 'orks.

/hat you need: A dish or tray of soil; a ırge clear glass jar; two similar ıermometers.

'hat you do: Lay the two thermometers ext to each other on the dish or tray of ›il. Cover one thermometer with the glass ır. The jar is our greenhouse. Stand the 'ay or dish in sunlight.

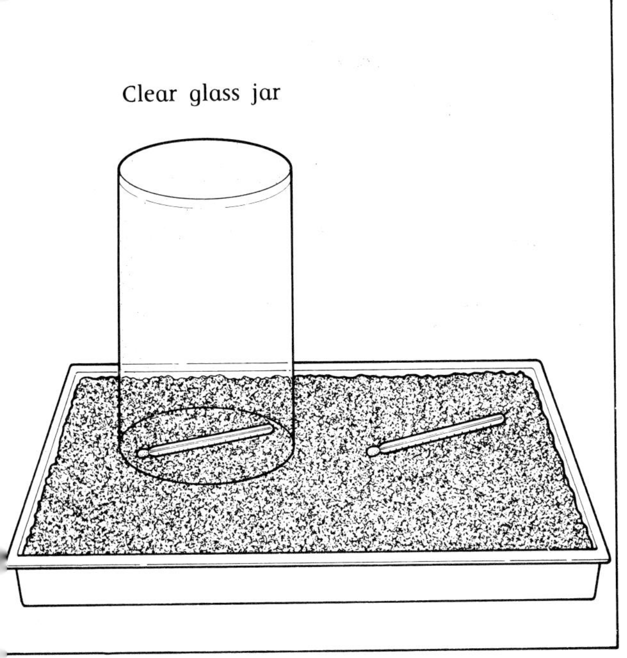

Read the temperature on the two hermometers every 30 minutes for an ıour or two. Which thermometer shows he highest temperature?

Now try the experiment with jars that ıre made of colored glass and also plastic ars. Which kind of greenhouse gives the ıighest temperatures?

7 How do different colors absorb the sun's rays?

What you need: Two similar thermometers; some sheets of paper, all the same size but different colors.

What you do: Lay the two thermometers near to each other in sunshine. What is the temperature?

Cover one thermometer with a sheet of black paper and the other with a sheet of white paper. Leave them in the sunshine for 15 to 30 minutes. Now quickly remove the two sheets of paper and read the temperature on each of the two thermometers. Under which sheet of paper was it the hottest? Under which sheet of paper was it the coolest? Which color paper absorbs most of the sun's rays?

Now try the experiment with sheets of paper of other colors. Also try a sheet of white rough paper and a sheet of white glossy paper.

What color clothes would it be best to wear if you were cold? What color clothes would you wear if you were hot?

If you cannot do this experiment in sunshine, you can always do it using the light and heat from a table lamp. See that the light falls equally on both sheets of paper, though.

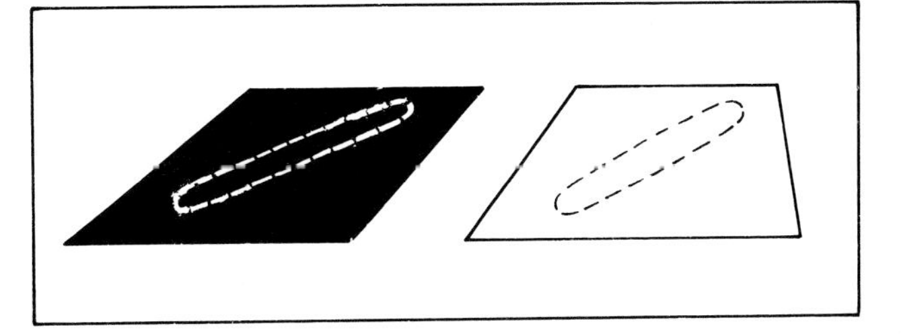

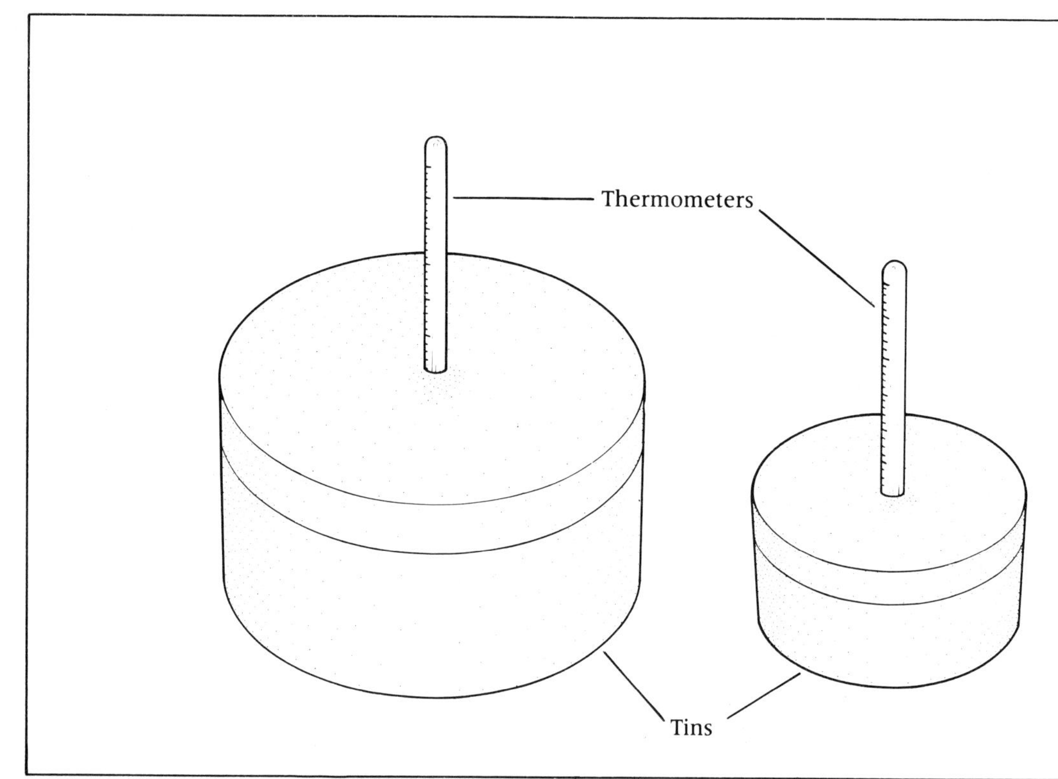

8 Do large things cool more quickly than smaller ones?

What you need: Two or more tins all the same shape but of different sizes. The tins must have lids. Some thermometers.

What you do: Make a hole in the lid of each tin. Make the hole big enough for a thermometer to go through.

Stand the tins in a row on the table. Take the lids off. Ask your teacher to fill each tin from a kettle of hot water. The water in each tin should be at the same temperature. Put the lid on each tin and put a thermometer through the hole in the lid.

Take the temperature of the water in each tin every 10 minutes. Make a graph to show how the water in each tin cooled. Use a different colored line for each tin.

Which tin cools the quickest? Which tin cools the slowest? Can you now understand why babies need to wear plenty of warm clothes, and why they wear more clothes in cold weather than grown-ups do?

Now do the experiment again. This time use two tins of the same size. Wrap cotton wool, fur, or a wool cloth around one of the tins. Leave the other tin as it is. Ask your teacher to put hot water in both tins. Take the temperature of the two tins every 10 minutes. Which cools slower? Which tin cools quicker?

Now try the experiment with other materials. Perhaps you could have one tin with cotton wool around it and the other with wool cloth around it. Which cools quicker? Which cools slower?

Freezing water

Many people say that hot water freezes faster than cold water. Is this true?

What you need: Two plastic cups or bowls (do not use glass!); a thermometer; the use of a freezer.

What you do: Fill one of the cups or bowls with hot water. Fill the other cup or bowl to the same level with cold water from the tap. Take the temperature of the two cups of water.

Put the two cups or bowls of water in a freezer. Look at them every 30 minutes. Does the hot water or the cold water freeze first. Is the saying true?

0 Melting ice cubes

What you need: Some ice cubes; three teacups; a spoon; a thermometer; a clock; cotton wool, cloth, aluminum foil, newspaper, and other materials.

What you do: What is the temperature of the air in your classroom? How long does an ice cube take to melt in your classroom?

Now wrap one ice cube in cotton wool, another in aluminum foil, another in cotton cloth, another in newspaper, and so on. See which ice cube melts first.

Now take the three cups. Put water at 50°F. in one cup, water at 68°F. in the next, and water at 86°F. in the third. Put one ice cube in each cup. Which ice cube melts first?

Does crushing an ice cube make it melt faster than one that has not been crushed?

Does sprinkling salt on an ice cube make it melt faster than one that has not had salt sprinkled on it?

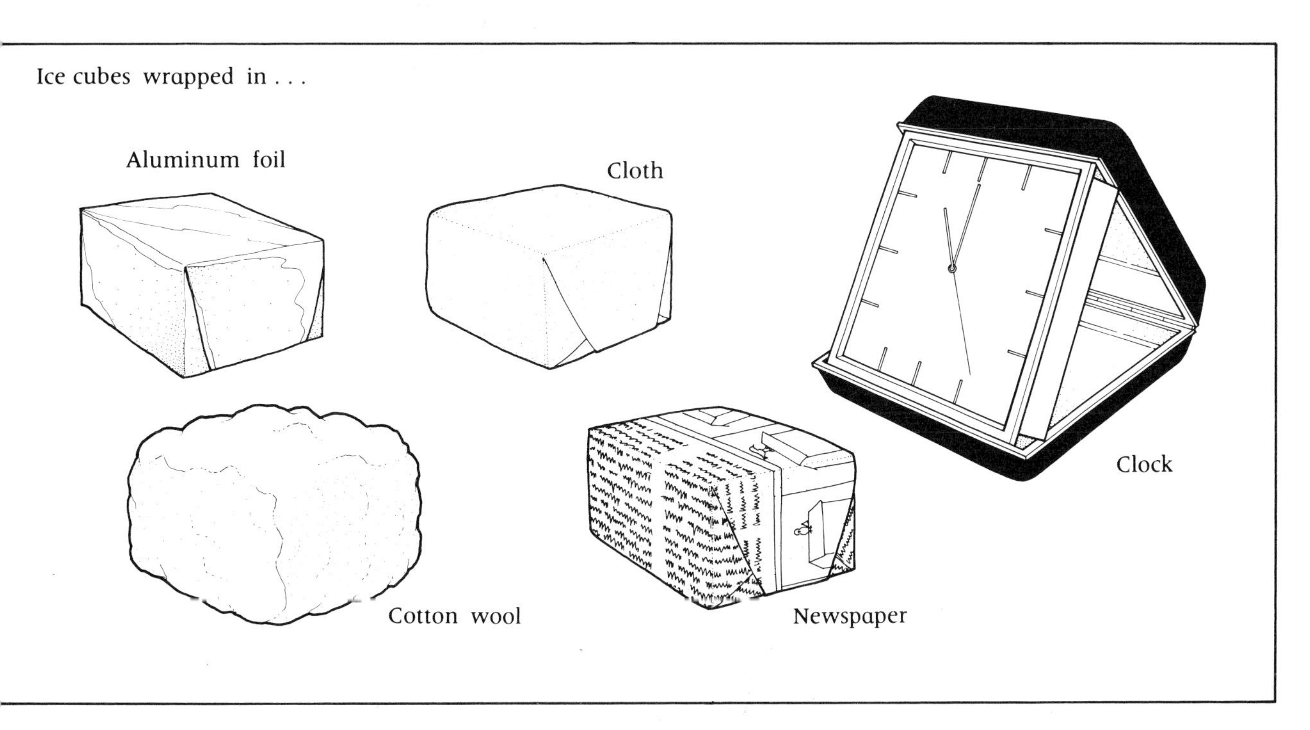

Glossary

Here are the meanings of some words that you might have met for the first time in this book.

Conductor: something that lets heat through it easily. Conductors usually feel cold to the touch.

Cold-blooded: When an animal's temperature changes with the temperature of the air or water around it, it is said to be cold-blooded.

Contract: when something gets smaller as it is cooled, it contracts.

Double glazed: a window with two sheets of glass with a layer of air in between.

Expand: when something gets bigger as it is heated, it expands.

Fuel: anything that will burn.

Insulator: something that does not let heat through it very quickly. Insulators feel warm to the touch.

Lava: the hot, liquid rock that comes out of a volcano.

Mammal: a warm-blooded animal whose body is covered with hair or fur. Mammals feed their young on milk.

Melt: when a solid turns into a liquid as it is heated, it melts.

Ore: a rock with a lot of metal in it.

Power plant: the large building where electricity is made.

Reflected: when rays of heat or light ar bounced back by something, we say they have been reflected.

Temperature: a measure of the hotness or coldness of something.

Thermometer: an instrument that is used for measuring temperature.

Warm-blooded: when an animal's temperature stays the same, whatever the weather, we say it is warm-blooded.

Acknowledgments

The publishers would like to thank the following for permission reproduce transparencies:

C. Alexander p. 9 (top), p. 16; Barnaby's p. 11 (top); Bruce Coleman p. 20 (left, center, right, bottom right and bottom left), p. 23 (center, left); Colorific, p. 7 (top); T Jennings p. 1 (top); Oxford Scientific Films p. 23 (bottom center); Picturepoint p. 4, p. 5 (left), p. 18 (bottom left), p. 20 (bottor right); M G Poulton p. 10, p. 18 (center right); Rentokil Ltd p. 9 (bottom); Shell p. 7 (bottom); A Souster p. 8 (top), p. 2((top right); Vision International: Anthea Sieveking p. 8 (center); ZEFA: H Lütticke p. 5 (right/Helbing p. 11 (bottom) K E Deckart p. 19/E M Bordis p. 19.

Illustrated by
John Barber, Norma Burgin, Karen Daws, Gary Hincks, Ed McLachlan, Chris Molan, Tony Morris, and Tudor Artists.

Index